Resilience & Friends: The Cycles of Nature

Written by Jessica Jane Robinson
Illustrated by Paige Mason
Design by Kara Greenfield

Resilience & Friends: The Cycles of Nature

By Jessica Jane Robinson

Copyright 2023 by Resilience Birthright Productions.

All rights reserved throughout the world. No part of this book may be reproduced in any form by any process whatsoever or by any means, electronic, or mechanical, including photocopying, recording, or any information storage or retrieval system without prior permission in writing from the publisher, except by a reviewer who wished to quote passages in connection with a newspaper, magazine, radio, blog or TV presentation.

This is a Resilience Birthright Productions book published by Resilience Birthright Productions

www.rbrorg.com

Print ISBN: 978-0-9997226-1-9

Written by Jessica Jane Robinson

Illustration by Paige Mason

Design by Kara Greenfield

Printed in USA

*This book is dedicated to the
loving community of friends and family
who have supported me throughout the years
and without whom
this work of art would not be possible.*

    There are many people I feel moved to thank who have motivated, inspired, and helped me create Resilience and Friends. I want to acknowledge my colleague, mentor, and friend, Ruth Abbe. She has been a beacon of light during my zero-waste career, and she has supported and encouraged me to implement my creative ideas and approaches even when they are "outside the box." I want to thank Paige Mason for her heartwarming illustrations in this book sharing and for the painting she made of me as Resilience years ago at a Comic Con. Paige's watercolor gift was the initial inspiration for me to create the spin-off Baby Resilience. Megan Zamora, my dear friend for almost two decades, has been my artist whisperer, and she has always supported my artistic visions and believed in me during my humble beginnings. Her feedback as a teacher inspired the lesson plans and my children's show, "Storytime with Resilience."
I thank Jen Abbe, Michelle Kuttner, and John Hatten for reviewing the work and providing feedback and helpful suggestions. Kara Greenfield has been my rock. I thank her for going beyond the call of duty with her help and skills. Kara's contributions and insight have been priceless.

    All the support helps make my work possible; I am only as good as my community and team.

Foreward

In 2020, the world came to a screeching halt. All the schools I supported and worked with for over a decade shut their doors. The teachers and principals I have known for years were all thrown online to teach the children with no preparation for any scenario or situation we experienced from 2020 to early 2021. I witnessed everyone scrambling to figure out how to teach the children under dire circumstances, and my green team parents desperately needed entertaining environmental education for their children online.

As a zero-waste advocate, problem solver, and passionate artist, I always seek ways to create and share meaningful content with my audiences. I recognized the need for more environmental educational content and sought grant funding to meet the demand.

"Resilience and Friends" is here to dissolve the barriers to preventing environmental education time in the schools and classrooms that existed even before 2020. The reason for the resistance towards the educational programs is that the academic standards teachers must teach yearly only allow room for learning within the set curriculum. Environmental, oceanic, and climate literacy have historically taken the backburner unless a passionate teacher or principal prioritized it. I want to solve this problem by removing the barrier and providing grade-specific educational content that can be easily integrated into the classroom, supporting the teachers to prepare students with 21st-century learning that reflects the world they live in.

I have written short stories called Resilience and Friends (RF). RF is an educational entertainment program targeting grades K-5. The content has a book of short stories, environmental education, lesson plans, and songs. RF lesson plans cover Next Generation Science Standards, Common Core Math, and English Language Arts Content Standards, which can be a tool for teaching in the classroom, online, and home-schooling.

This book, in particular, targets grades K-2. I will be publishing a second book that targets grades 3-5 in the next year or so (keep a lookout for book #2 in 2024/2025).

The following table is a list of lesson plans meant to support the content of this book through environmental education. Resilience & Friends lesson plans cover Next Generation Science Standards, Common Core Math, and English Language Arts Content Standards, which can be a tool for teaching in the classroom, online, and home-schooling.

Grade Code

Lesson Number	Subject/Title	Cirriculum	Category	Kindergarten	First Grade	Second Grade
1.1	Reading Comprehension	Common Core	ELA - Reading: Literature	CCSS.ELA-LITERACY.RL.K.3	CCSS.ELA-LITERACY.RL.1.2	CCSS.ELA-LITERACY.RL.2.7
		Common Core	ELA - Language Vocabulary	CCSS.ELA-LITERACY.L.K.4.A	CCSS.ELA-LITERACY.L.1.5	CCSS.ELA-LITERACY.L.2.5
1.2	Environmental Science Flashcards	Common Core	ELA - Language: Vocabulary	CCSS.ELA-LITERACY.L.K.4.A	CCSS.ELA-LITERACY.L.1.5	CCSS.ELA-LITERACY.L.2.5
		Common Core	ELA - Language: Vocabulary	CCSS.ELA-LITERACY.L.K.5.C	CCSS.ELA-LITERACY.L.1.5.C	CCSS.ELA-LITERACY.L.2.5.A
2.1	How Do Apples Grow?	Common Core	ELA - Reading: Literature			CCSS.ELA-LITERACY.RL.2.7
		Common Core	ELA - Reading: Literature			CCSS.ELA-LITERACY.RL.2.10
		Next Gen Science	Life Sciences		1-LS1-1 From Molecules to Organisms: Structures and Processes	
		Next Gen Science	Life Sciences		1-LS3-1 Heredity: Inheritance and Variation of Traits	
2.2	Plant Life Cycle	Next Gen Science	Life Sciences	K-LS1-1 From Molecules to Organisms: Structures and Processes		2-LS2-1 Ecosystems: Interactions, Energy, and Dynamics

Download Lesson Plans

Use the QR code or follow the link below to access and download lesson plans for *Resilience & Friends: The Cycles of Nature*.

https://1.rbrorg.com/LessonPlans

				Grade Code		
Lesson Number	Subject/Title	Cirriculum	Category	Kindergarten	First Grade	Second Grade
2.3	Pollinator Timeline	Common Core	ELA - Reading: Literature			CCSS.ELA-LITERACY.RL.2.7
		Common Core	ELA - Reading: Literature			CCSS.ELA-LITERACY.RL.2.10
		Next Gen Science	Life Sciences			2-LS2-2 Ecosystems: Interactions, Energy, and Dynamics
3.1	Earth & Human Activity	Next Gen Science	Earth & Space Sciences	K-ESS3-3 Earth and Human Activity		
3.2	Counting with Baby Res4	Common Core	Math - Measurement & Data	CCSS.MATH.CONTENT.K.MD.B.3		
		Common Core	Math - Operations & Algebraic Thinking		CCSS.MATH.CONTENT.1.OA.D.7	CCSS.MATH.CONTENT.2.OA.A.1
		Common Core	Math - Operations & Algebraic Thinking		CCSS.MATH.CONTENT.1.OA.D.8	
4.0	Differences & Similarities	Next Gen Science	Life Sciences	K-LS1-1 From Molecules to Organisms: Structures and Processes		
		Next Gen Science	Life Sciences	K-ESS3-1 Earth and Human Activity		

Table of Contents

Chapter 1: Terry the Apple Tree 11

Chapter 2: Pollinator Friends 21

Chapter 3: Someone's Treat, Another's Compost Story 35

Chapter 4: Superheroes Save Trees: Recycle, Reduce, Reuse 49

 "Storytime with Resilience" (STR) is online environmental education for teachers and parents to leverage for learning time with their children while utilizing storybooks and lesson plans.

 Founder Jessica Jane Robinson plays Superhero Resilience as she brings the stories of Resilience & Friends to life for K-5 audiences. Resilience takes her audience on investigative journeys outside of her story room into nature, where she and her audience can explore concepts they are learning about in more depth. Each episode covers essential topics that give children information about the wonders of the planet, natural habitats, how everything is interconnected, and why zero waste practices are so important.

https://1.rbrorg.com/ResStorytime

or

https://www.youtube.com/@resilienceandfriends

Chapter 1: Terry the Apple Tree

One autumn day, Baby Res finds herself eating an apple under Terry the Apple Tree's branches.

She looks at the apple and wonders.

Baby Res looks up at Terry the Apple Tree and asks, "Terry the Apple Tree, how *do* apples grow?"

Terry the Apple Tree smiles and says, "An apple's story starts with the little apple seeds inside the core of your apple."

Terry the Apple Tree continues, "First, plant the apple seed in **soil** with nutritious **compost** on top."

Terry the Apple Tree shares happily, "The seed likes **sunlight** and **water**. The seed grows best outside where there's fresh **air**. Nature gives the seed all it needs to grow."

Terry the Apple Tree continues, "The **seed** will sprout into a **seedling** and grow roots deep into the soil.

After two years, the seedling will turn into a **sapling**, and in a few more years, it will grow into a tree.

The tree will grow **flowers** in the spring."

Excited about the story, Baby Res looks at her apple and asks,

"But how do apples *grow*?"

Mr. and Mrs. Wiggles appear from the ground and greet everyone, "Hello."

Sunny the Squirrel appears from Terry the Apple Tree's branches, "Hello, friends."

Baby Res greets her friends and sits, and she thinks. Baby Res has another question; "Friends, I still don't know; how apples grow?"

Just then, Beau the Butterfly and Bumbles the Bee flutter by and ask together, "The apple tree's flowers turn into an apple. Can we tell this story of the great pollinators?"

They all nod their heads and say, "Yes. We love story time."

Beau the Butterfly and Bumbles the Bee say together, "Baby Res, we will tell you how apples grow."

Chapter 2: Pollinator Friends

Baby Res, and her friends all sit under Terry the Apple Tree's branches.

Beau the Butterfly and Bumbles the Bee begin their story of the great pollinators.

Beau the Butterfly flaps his wings and flutters around,

"**Pollinators** are important for plants to grow."

Beau the Butterfly continues, "Butterflies move from flower to flower.

We spread pollen from one flower to the next flower as we flutter around.

This is called **pollination**."

Beau the Butterfly shares, "We like the flowers of vegetables and herbs. These are some of our favorites."

Parsnips

Artichokes

Lettuce

Celery

Broccoli

Can you read them all?

Lavender

Peas

Rosemary

Beau the Butterfly flutters with excitement, "Butterflies help these plants make seeds to grow more plants; it's our job as pollinators!"

Baby Res asks, "Pollination is a job?"

Beau the Butterfly and Bumbles the Bee both smile and reply, "Yes."

Beau the Butterfly continues, "Our reward is…."

Bumbles the Bee finishes Beau's sentence, **"Nectar."**

Baby Res looks at her apple, "I still don't know how a flower can turn into an apple!"

Bumbles the Bee buzzes her wings excitedly,

"This is my part of the story. When I land on an apple tree's flowers, called blossoms, I pick up pollen on my feet and body while I collect delicious nectar."

Bumbles the Bee continues, "Then I buzz to another apple tree to collect more nectar from more flower blossoms.

Each time I visit a blossom, some pollen falls off my body onto the **stigma** of another flower.

Once this happens, I pollinated the flower. The petals fall off, and then the apple starts to grow."

Bumbles the Bee says with excitement, "After I have enough nectar in my sacks, I buzz all the way home to my hive, and my sisters help me turn my nectar into honey."

Baby Res happily claps and pulls out her glass honey jar, "I love to eat honey with my almond butter sandwiches!"

Bumbles the Bee and Beau the Butterfly say together, "Pollinators are important for plants to grow."

Baby Res finishes her apple and holds up her apple core and says, "Wow, what a story!"

Mr. and Mrs. Wiggles smile and look at Baby Res's apple core and ask, "May we have your apple core? We want to eat it for dinner and make **compost** for Terry the Apple Tree."

Baby Res smiles and happily gives Mr. and Mrs. Wiggles her apple core and asks, "What is compost?"

Mr. Wiggles asks the group cheerfully, "Who wants to hear another story?"

Baby Res and her friends reply, "Me!!!!"

Chapter 3: Someone's Treat, Another's Compost Story

Baby Res and her friends sit quietly by the picnic area under Terry the Apple Tree, waiting for Mr. and Mrs. Wiggles to share their story about compost.

Mrs. Wiggles starts the story, "Well, my friends, Mr. Wiggles loves fruit, and apple cores are one of his favorites."

Mr. Wiggles continues to share, "I also like strawberries, peaches, apricots, banana peels, and all types of melon! I sure like to eat!

Mrs. Wiggles enjoys many vegetables; I am the sweet tooth in the family."

Mr. Wiggles grins and says, "Baby Res, if you ever have leftover fruits and vegetables, please give them to us; we'll recycle your food scraps."

Mrs. Wiggles smiles and says, "Mr. Wiggles sure likes to eat."

Baby Res questions, "You'll recycle my food scraps?! How?!"

Mrs. Wiggles answers, "When we eat the leftover food scraps, the food will pass through our body and exit from our tail. We turn food scraps into a soil called worm compost."

Terry the Apple Tree joins the story, "I love Mr. and Mrs. Wiggles' compost! It's so healthy for the soil I grow in.

The compost helps my roots grow strong, deep into the ground, so that I can produce more flowers and apples.

I even release more **oxygen** into the air with compost!"

Mr. Wiggles continues, "When people throw food scraps away into the landfill, it creates a gas called **methane**, harming our environment.

But when we compost, we stop methane gas, create nutritious soil that traps in gases, and help Terry the Apple Tree clean our air."

Baby Res asks, "Terry the Apple Tree, how do you clean our air?"

Terry the Apple Tree explains, "I can clean our air by a process called **photosynthesis**. I absorb sunlight and the gases you breathe out, called carbon, through my leaves.

I absorb water from the ground, and compost helps the soil stay moist with plenty of water, and then I release oxygen from my leaves back into the air that you all breathe."

Baby Res is delighted, "Wow, Beau the Butterfly, Bumbles the Bee, Terry the Apple Tree, Mr. and Mrs. Wiggles, you are all my heroes. You help our planet every day!"

Mr. Wiggles says, "Well, being a hero is easy when you get to eat."

Everyone laughs and says, "Mr. Wiggles sure likes to eat."

Sunny the Squirrel jumps up from his seat, excited to share next, "Baby Res and all my friends, would you like to hear one more story of how squirrels are heroes too?"

Everyone looks at Sunny the Squirrel and smiles. Baby Res claps her hands, "Yes, another story, please!"

Chapter 4: Superheroes Save Trees: Recycle, Reduce, Reuse

Sunny the Squirrel scurries up Terry the Apple Tree's trunk, staring at his friends, and begins the story, "Well, friends, I'm a gray squirrel, and we're known as the great forest regenerators! My ancestors and I are the planet's heroes."

Baby Res is excited, "You're a hero?"

Sunny the Squirrel says with sadness, "Humans have cut down many of our trees, which are the homes, shelter, and food for animals in the forest."

Baby Res thinks of paper, "Sunny the Squirrel, humans use trees for paper, to build homes, furniture, and other things."

Sunny the Squirrel continues, "The forest has been damaged from all the trees cut down, and the animal kingdom and my family need the trees because they are our home. We are heroes because we collect seeds and nuts in the autumn. We first eat the seeds and nuts that are rotting or infested with bugs."

With disgust, Baby Res says, "Yuck, you eat bugs?!"

Sunny the Squirrel smiles and says, "Yes, insects and larvae are delicious! Squirrels also like flowers, buds, and bark in the late winter. In the spring, we eat fungi and fruits and berries in the summer. In the fall, we like seeds and nuts."

Sunny the Squirrel continues, "We can't eat all the seeds and nuts we bury; they sprout and become brand-new trees in the forest.

The gray squirrel is a forest regenerator.

We help plant and grow trees, making us heroes of the forest!"

Baby Res looks at her friends, "I am a human, but I can be a hero too!"

Baby Res holds up her glass honey jar, "When this is empty, I can reuse this glass jar, and when I don't need it anymore, I can recycle it."

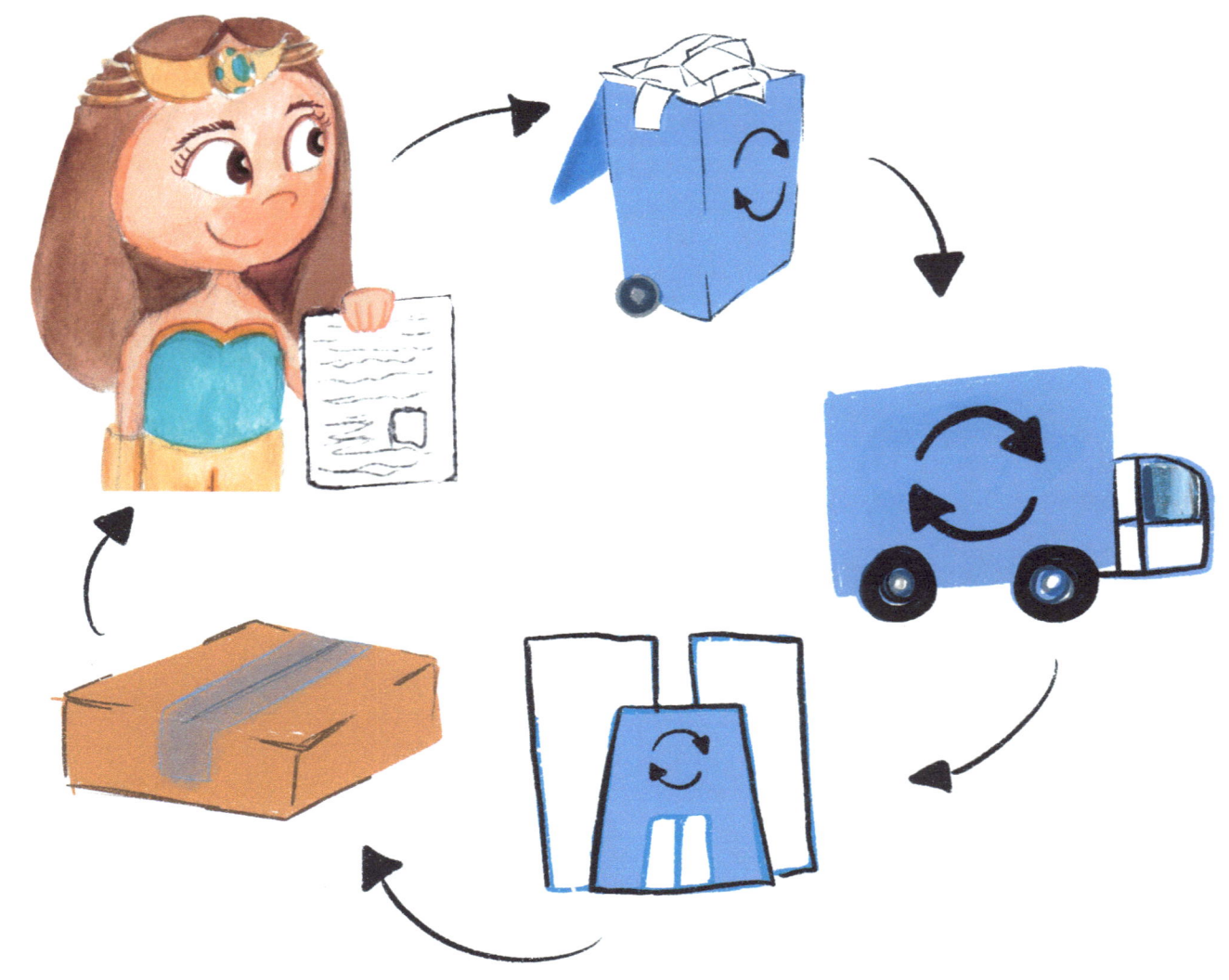

Baby Res continues, "At home, I recycle all my paper. Paper comes from trees, so we don't need to cut down trees when recycling paper. Recycling makes new paper! I am a hero, too."

Baby Res happily shares, "When my family cooks, I recycle all the metal cans and aluminum foil. Metal comes from minerals found in the earth and even mountains. So when I recycle metal, the metal turns into new metal cans, and I help the planet!"

Baby Res says proudly, "I recycle all my plastic containers too!

But even better, I use my reusable water bottle, utensils, lunch box, containers, and napkin every day.

I practice refusing single-use plastics daily because I know that is a way I can be a hero, too!"

All the friends smile at Baby Res and say, "Yes, we are all heroes!"

Baby Res continues, "Together, we can protect our planet and keep our homes happy and clean."

Baby Res holds one apple seed she saved from her apple, "I still have an apple seed left."

Sunny the Squirrel jumps excitedly, "Baby Res, let's plant the tiny apple seed. Let's go plant a tree!"

Mr. and Mrs. Wiggles join in, "We'll bring the compost."

Discover more Resilience Birthright content! Follow the QR code or link below to access official Resilience Birthright website.

rbrorg.com

Join our newsletter to stay updated on new content including the release of new books, lesson plans, and other forms of media related to zero-waste education. Use the QR code or link below to sign-up for free!

https://1.rbrorg.com/NewsLetter

or

https://mailchi.mp/resiliencebirthright/mlcavj98qo

Follow Resilience & Friends on Instagram!

www.ingramcontent.com/pod-product-compliance
Lightning Source LLC
Chambersburg PA
CBHW050855010526
44118CB00005BA/177